AMONG THE ASPEN

Half-title: Late-summer aspens,
Dallas Divide, western Colorado.
Pages x-xi: Early snowfall backdrops autumn gold aspens,
San Juan National Forest, southwestern Colorado.

For information, address Northland Publishing Company,
Post Office Box N, Flagstaff, Arizona 86002.
First Edition
ISBN 0-87358-521-6 (softcover)
ISBN 0-87358-522-4 (hardcover)

Library of Congress Catalog Card Number
91-2791
Cataloging-in-Publication Data
Petersen, David.
Among the Aspen / text by David Petersen ;
photographs by Branson Reynolds. -- 1st ed.
160 pp.
ISBN 0-87358-522-4 (hc) : $35.00 (tentative).
-- ISBN 0-87358-521-6 (sc) : $24.95 (tentative)
1. Natural history--Colorado. 2. Populus tremuloides--
Colorado--Ecology. 3. Forest ecology--Colorado. I. Title.
QH105.C6P47 1991
508.788--dc20   91-2791
CIP

Composed in the United States of America
Printed in Hong Kong by Lammar Press

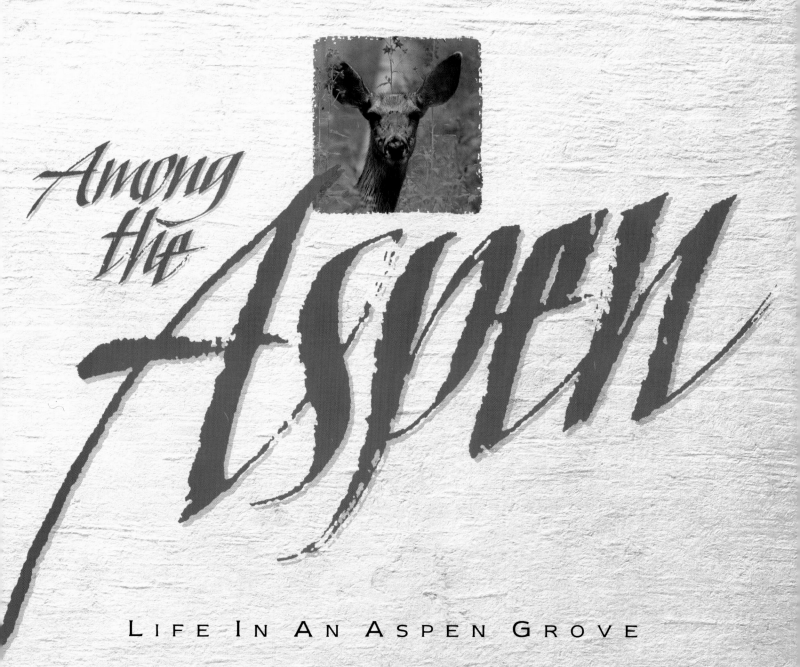

Among the Aspen

LIFE IN AN ASPEN GROVE

TEXT BY

DAVID PETERSEN

PHOTOGRAPHS BY

BRANSON REYNOLDS

NORTHLAND PUBLISHING COMPANY

For Christine

CONTENTS

ONCE UPON A TIME, when the Seven Wise Men of Greece were met together at Athens, it was proposed that each of them should state to the others what he considered to be the greatest wonder in the Creation. One of them asserted that nothing was so wonderful as the heavenly bodies. He explained the opinions of some of the astronomers respecting the fixed stars, that they were so many suns, each having planets rolling round them, which were stocked with plants and animals like this earth. ❦ Fired with the idea, they instantly agreed to supplicate Jupiter that he would at least permit them to take a journey to the Moon, and remain there three days, in order that they might view the wonders of that place, and give an account of them to the world at their return. Jupiter consented: he directed them to assemble on a high mountain, where a cloud should be in readiness to convey them thither. They did so, and took with them some men of talents, to assist in describing and painting the objects they should discover. At length they arrived at the Moon, where they found a palace fitted up for their reception. ❦ On the day after their arrival, they were so much fatigued with their journey, that they remained in the house till noon; and, continuing still faint, they partook of a delicious entertainment, which they relished so much that it quite overcame their curiosity. This day they only saw, through the windows, a delightful country, adorned with luxuriant verdure, and with flowers exquisitely beautiful, and heard the melodious singing of the birds. ❦ The second day they rose very early, to commence their observations; but some elegant females of the country, calling upon them, advised that they should first recruit their strength before they exposed themselves to the laborious task they were about to undertake. The sumptuous banquet, the rich wines, and the beauty of these females, prevailed over the reso-lution of the strangers. Music was introduced, the young ones began to dance, and all was turned to jollity; so that the whole of this day seemed dedicated to gaiety and mirth, till some of the neighbors, envious of their happiness, rushed into the room with swords. With some difficulty these were secured; and it was promised, as a recompense to the younger part of the company, that, on the following morning, they should be brought to justice. ❦ On the third day the trial was heard; and, in consequence of the time occupied by the accusations, pleading, exceptions, and the judgment itself, the whole day was occupied, and the term which Jupiter had allowed to the Wise Men expired. ❦ When they returned to Greece, the whole country flocked around them to hear the wonders of the Moon described; all they could say, for it was all they knew, was this: that the ground was covered with verdure, intermixed with flowers; and that the birds sang delightfully among the trees; but what was the nature of the flowers they had seen, of the birds they had heard, or of the country they had visited, they were entirely ignorant. On which they were every where treated with contempt. ❦

This fable was applied, by Linnaeus [Swedish botanist Carl von Linnaeus, 1707-78], to mankind in general. In youth we are too feeble to examine the great objects around us; all that season, therefore, is lost amidst weakness, indolence, luxury, and amusement. We are little better in manhood; settling ourselves in life; marrying; bustling through the world; overwhelmed, at length, with business, cares, and perplexities, we suffer those years also to glide away. Old age succeeds; still some employments intervene, till, at last, we have passed through the world, without scarcely a single recurrence to the admirable works of our Creator; and, in too many instances, even without having duly considered the end for which we were brought into it. This, with a few exceptions, is the progress of man through life. It is true that no one is able to avoid being led, by his own feelings, occasionally to notice the wonderful productions with which he is surrounded. All can remark the beautiful verdure of the fields and woods; the elegance of the flowers; and the melodious singing of the birds; yet few indeed give themselves the trouble of proceeding a single step further, or exhibit any desire of examining into the nature of these astonishing combinations.

—Reverend W. Bingley, A.M.

from *The History of Animals*, circa 1825

ACKNOWLEDGMENTS

FOR THEIR UNFLAGGING ENCOURAGEMENT, friendship, and direct help in this and other such projects, I wish to thank Carolyn, my live-in editor; A.B. Guthrie, Jr., my exemplar and coach; Betti Albrecht, my talented Northland editor; and, as always, Ed, wherever he may be, whatever he may be up to — *Vox Clamantis in Deserto.*

FOR TEN YEARS AND MORE NOW, my wife, Carolyn, and I have lived in a tiny board-and-batten cabin squatting on an equally tiny acreage tucked up in a far back corner of one of the oldest, most neglected, and least populated summer-home subdivisions in south-western Colorado. Elevation: exactly eight thousand feet above the distant western sea. The dirt subdivision roads are worse than bad — like bumping up a dry rocky streambed — and we like it that way; the ruts and bumps repel real estate agents and lot-shoppers alike; maximum privacy for minimum investment. It's the best (meaning the most private and remote) we can manage just now, and plenty good enough.

Sheltering our little mountain redoubt is an island grove of mixed-maturity quaking aspens, several hundred in number. The San Juan National Forest lies a most pleasant (if somewhat steep) ten-minute walk to the east and north, the Weminuche Wilderness boundary a half-day's hike beyond.

Unwelcome visitors (human most always) are rarely a problem, for, in addition to the roads up to our place being steep and rutted, they are closed completely to wheeled access when the serious snows arrive, as they generally do, toward the end of each November.

Welcome visitors to our little aspen haven include mule deer, elk, the occasional nocturnal black bear, plus a multitude of smaller animals and birds — all of them dependent, to one degree or another, on the aspen ecology for their welfare.

This book is a biography of that ecology — the trees themselves, of course, but also the dynamic universe of plants and animals that look to the aspen grove for physical and spiritual nurture. Among this number, I have come to count myself. Living here has enlarged and redirected my life. *Among the Aspen*, then, is a personal natural history.

My partner in this interpretive venture is photographer Branson Reynolds, a close friend and keen observer and recorder of nature, who, with less than a half-mile of forest and wrinkled rocky earth separating his cabin from mine, is one of our nearest year-round neighbors up here among the aspens. Branson's sterling images, not my struggling monologues, make this book. ❦

Heaven
on
Earth

1

HEAVEN ON EARTH

The clearest way into the universe
is through a forest wilderness.

— JOHN MUIR

I remember it like a dream.

I'm standing alone in a parklike grove of quaking aspens. I have come here, as on so many evenings before, to see what I can see. Wishing not to disturb the sylvan tranquillity of this special place (only just one of many I have discovered and laid emotional claim to hereabouts, but special nonetheless), I am dressed in camouflage and attempt to keep my movements as imperceptible as the progress of time, as quiet as — what? Well, as quiet as a bumbling, balding, middle-aged human animal can manage.

Just ahead, a predictably ill-tempered gray jay, a large, white-crowned, black-tipped gray bird (*Perisoreus canadensis*), also called Canada jay, Whiskey Jack, and — the most accurate of all its many nicknames, camp robber — screams its harsh familiar curse, then drops from its lofty perch, strokes twice with noisy wings, and sails away, tilting low amongst the stark, upright beams of the quaking aspen forest.

This minor commotion in turn startles to flight a pair of tiny sweet chickadees. (Chickadee: A most beautifully onomatopoetic name it is, given the little gray-and-white bird's distinct, repetitive, all's-well call of *chicka-dee-dee-dee*.) Up the valley a ways — I can't judge just how far —

BUGLES OF RUTTING BULL ELK HARMONIZE WITH RUSTLING AUTUMN ASPEN LEAVES.

an autumn-enamored bull elk issues a drawn-out bugle…like a bent high note on a saxophone.

I move slowly on, following a dim but familiar game trail that twists and dodges through the aspens, hoping to catch a glimpse of the bull. Stepping carefully, I stop often to look, listen, test the downslope evening breeze for familiar wild scents: the tangy, pungent smell of autumn aspens; the heavy barnyard funk of elk in rut; the fresh, willowy aroma of the small spring pool that waters this tight little valley with its jungle of aspen-evergreen and lush understory of pine drops the color of dried blood, lady ferns both common and alpine, giant larkspur and cow parsnips, angelica, chokecherry, bearberry, wild raspberry, and other lush living things whose names I have yet to learn and perhaps never will. The going here is quiet, easy, though the undergrowth is so profuse that I can see no more than a few yards in any direction around me.

Suddenly, from out of a tangle of Gambel oak just ahead, an animal appears: roundish, the color of a Hershey bar, furry and fat, the size (more or less) of a badger.

What it is, I realize with a start, is a bear cub, small for this late in the season, but a bear cub nonetheless. I am surprised, elated — but also a little confused.

Before I have time to sort out my thoughts, a second cub comes

---

"…stop often to look, listen, and test the downslope evening breeze for familiar wild scents…"

---

shambling out of the same island of scrub oak. And a third. And all three are headed straight for me.

Notwithstanding the considerable time I spend alone and quiet in the woods, and even given the generous black bear population hereabouts, my close-up bruin sightings over the past decade have been, as they say, few and far between. The appearance of this wee chocolate trio is great good luck. Still, I am distracted by a couple of curiosities: *Where might be the mother?* And, perhaps a more pressing concern, *how big is she?*

In the Beginning

2

IN THE BEGINNING

Has joy any survival value in the operation of evolution?
I suspect that it does....Where there is no joy there can be no
courage; and without courage all other virtues are useless.

— EDWARD ABBEY
from *Desert Solitaire*

The fossil record tells us that aspens have adorned the
beautifully rugged landscapes of western North America
since midway through the Miocene epoch, suggesting a tenure of
perhaps fifteen million years. Those that grew then, like those that live
today, are classified taxonomically within the genus *Populus*. Other
common names for the quaking aspen include popple (antiquated;
you'll run into it occasionally in older texts), trembling aspen (from the
scientific name *Populus tremuloides;* a variation on "quaking"), and
golden aspen (from the high yellow of the leaves in fall).

Of the world's four aspen species, the only North American native
other than the quakie is the eastern bigtooth (*P. grandidentata*), which
is thought to be closely related to the ancestral fossil varieties of
Miocene vintage. Contemporary bigtooth and quaking aspens, where
their ranges overlap, readily hybridize. Likewise, the fossil record
suggests that interbreeding was common. Hence, it's probable that only
their conflicting tastes in environments — a conflict that minimizes
overlap of the two species — has allowed the trembling and bigtooth

OPPOSITE: ASPEN SAPLINGS DEVELOP
WITHIN THE ROOT LAYER OF AND SPROUT
DIRECTLY FROM THE MATURE PARENT
TREE.

PRECEDING PAGES: RED-LEAFED ASPEN
CLONES, BETWEEN COAL BANK HILL AND
MOLAS DIVIDE, SOUTHWESTERN
COLORADO.

aspens to maintain distinct identities. Here in North America, the quaking aspen is not only the most widespread of native trees, but in many areas the *only* upland hardwood.

Looking a bit farther up the family tree, the aspens, as members of the poplar genus, consequently belong to the willow family (Salicacaea). The aspen's closest living relatives are the cottonwoods and poplars. In fact, young cottonwoods, with their white bark and loose-jointed leaves

that dance and rattle in the breeze, bear some resemblance to their cousins the aspens; likewise, older aspens, if your view is confined to the gray, deeply furrowed bark of their lower trunks, resemble cottonwoods. The relationship is apparent.

Surprisingly, the birches (genus *Betula*) —one species of which, the paper birch (*Betula papyrifera*, also called canoe birch or white birch), with its tall, slender trunk, white bark, and high crown of lime-green leaves — can easily be mistaken for quaking aspen, but in fact is unrelated. Adding to the confusion, throughout much of Canada and portions of the northwestern United States, the ranges of quaking aspens and paper birches overlap.

While the quaking aspen itself has in the past been subdivided into as many as a baker's dozen subspecies or races, current taxonomy con-

> "…aspens have adorned North America since midway through the Miocene epoch, suggesting a tenure of perhaps fifteen million years."

ABOVE: AUTUMN ASPEN LEAVES AFLOAT IN STONE EROSION POCKET.

OPPOSITE: LICHEN AND ASPEN LEAF REVEAL NATURE'S DELICATE BEAUTY.

siders it a single, albeit highly variable, form across its entire North American range.

The northern range of the quaking aspen in North America includes most of Alaska and Canada. In the eastern lower forty-eight,

the quakie occurs from New England as far south as Virginia. The tree is prolific throughout the Great Lakes states, spotty but widespread across the northern Midwest, and a prized institution along almost the entire length of the Rocky Mountains — from Canada on the north, as far south as New Mexico and Arizona. Even below the border, down in Baja California and the northern states of Mexico, rare cool stands of hardy aspen persist here and there.

The two states having the greatest concentrations of quaking aspens — more than a million acres each — are Colorado (primarily

"...the higher and farther north aspen occur, the stronger becomes their preference for sun-warmed southern slopes."

on the Western Slope) and Utah (right down the middle). Further, with nearly two-thirds of all western aspen growing on public lands, the quakie is among the most democratic of America's trees.

In the Rockies, the quaking aspen prefers, though is not restricted to, subalpine slopes where it intermingles with pines in the lower reaches and spruce and fir higher up. Here in my southwestern Colorado stamping grounds, as throughout the southern Rockies, quakies thrive in the sixty-five-hundred- to nine-thousand-foot elevation zone of the ponderosa pine, thinning out and eventually disappearing higher up as spruce and fir replace pine. At the extremes of its elevation tolerance, the quaking aspen occurs from sea level up to twelve thousand feet.

It is interesting, though hardly surprising, that the higher and farther north aspen occur, the stronger becomes their preference for sun-warmed southern slopes. While the quakie will invest in almost any directional aspect across the middle latitudes and elevations of its Rocky Mountain range, it prefers the cooler, breezier, north-facing slopes down in the southern extremes of Arizona, New Mexico, and Mexico.

By and large, the quakie is partial to areas that offer lengthy growing seasons punctuated by cold winters. In semiarid climes, it generally seeks out pockets of moist soil such as stream courses and valleys.

THE ASPEN LEAVES' CAROTENOID PIGMENTS ARE REVEALED EACH FALL, PROVIDING A STARK COUNTERPOINT TO THE GREEN CONIFERS.

AMONG THE ASPEN

An eager invader of open, sunny slopes and meadows, the quakie
is a post-holocaust opportunist, among the first trees to spring up in
burned-over areas. If fire recurs now and again — every few decades —
to cull the old and ill and make way for the young and eager, an aspen
grove can maintain dominance in a given area for hundreds, even
thousands, of years. Without fire, a quakie grove tends to die out or
move slowly on, giving way to the slower-
growing but hardier and much longer-
lived conifers.

The undulating southwest slope of my
little Colorado acreage is a typical example
of this progression.

Tree-ring studies show that in recent
historical times — that is, before Europeans
came along to disrupt the natural forest
cycle with logging, field-clearing, and fire
suppression — wildfires occurred on an
average of every seven to ten years throughout the ponderosa-aspen forests of the San
Juan Basin. Something less than a century
ago, one of these lightning-caused blazes
swept across the hillside where I now live. The flames of this particular
fire burned hot enough to kill all but a relative handful of the largest,
most ancient ponderosa pines — thick-barked, high-crowned trees
tough enough to withstand all but the most intense and prolonged

ABOVE: WILDFIRE PLAYS A NATURAL AND
IMPORTANT ROLE IN THE REGENERATION
OF ASPEN FORESTS.

OPPOSITE: STANDING TALL AND STRAIGHT,
EAGER YOUNG ASPENS GROW STEADILY
SKYWARD.

PRECEDING PAGES: WITHOUT FIRE, A
QUAKIE GROVE TENDS TO DIE OUT OR
MOVE SLOWLY ON, GIVING WAY TO THE
SLOWER GROWING BUT HARDIER AND
MUCH LONGER-LIVED CONIFERS.

"...the scorched earth erupted with vigorous
new quaking aspen saplings."

conflagrations. A few charred stumps remain as mute evidence of those
less hardy and fortunate.

Soon after, the scorched earth — warmed and stimulated by sunlight
previously blocked by a thick overstory of conifer foliage, and fertilized
with nutrients released by flame from the burned vegetation — would
have erupted with vigorous new grasses, forbs, Gambel oak, choke-

34

cherry, and serviceberry...and quaking aspen saplings. Hundreds, maybe thousands of them.

Within a couple of seasons, willowy young aspens must have bristled like an old man's whiskers across the upper portion of the blackened and balded slope, while scrubby Gambel oak and berry brush began filling in down lower. A Rocky Mountain aspen grove was born (or, perhaps and even likely, reborn).

With a maturity rate that's rapid for trees, roughly paralleling that of humans, the aspens would have attained young adulthood in just two or three decades, gaining more or less a foot of new height per year. Today, maybe seventy or eighty years since the fire that started it all, the trunks of the largest quakies in this grove average eighteen inches dbh (a forester's term for diameter at breast height), and stand sixty feet or taller.

(For the record: Over in Utah, a harsh, arid state populated by tough survivors of every stripe, ancient aspens live today having trunk diameters of fifty-four inches dbh, and heights of one hundred twenty feet.)

Sadly, saddled with an average life span roughly equivalent to yours and mine (if we're careful and lucky), most quakies can expect maximum tenures of only eighty to a hundred years on this lovely earth. Only the rare, tough, or fortunate monarch will live to enjoy two centuries of springs, and most quakies die young, the commonest killer being a nominally humanlike ailment — heart rot.

Hence, today, most of the oldest aspens, highest up the slope of my little grove, have already succumbed, being replaced by an insidious encroachment of pine, spruce, and fir, all of which need the protective shade provided by the aspens' high, domed foliage crowns to nurse them through the delicate seedling stage. The middle-aged aspens a little farther down the hill, amongst which squats my cabin, still thrive — though we lose a few almost every year from various causes, and conifer seedlings are gaining a firm foothold within their territory. Farthest down the hill, long the domain of oakbrush, grasses, and wildflowers, aspen saplings proliferate.

In short, our little quakie stand is sliding slowly downhill. This pattern is a blueprint for the lives of many mountain aspen groves: to invade a burned-over area from above and migrate gradually downslope, the abandoned upper territory filling in with evergreens.

ABOVE: MULE DEER BUCK AMONGST SUMMER WILDFLOWERS.

OPPOSITE: MOUNTAIN WILDFLOWERS, LIKE THESE PURPLE ASTERS, YARROW, AND CONEFLOWERS, ARE IMPORTANT AND COLORFUL MEMBERS OF THE ASPEN FOREST COMMUNITY.

FOLLOWING PAGES: SPRINGTIME DANDELIONS AND ASPENS WITH THE NEEDLE MOUNTAINS IN THE BACKGROUND, SOUTHWESTERN COLORADO.

It's a joy living here among the aspens. Joyful in a great many ways. During all the years I lived in cities (I was, much to my adolescent dismay, born and raised in a bustling capital city out on the Great Plains), I, like everyone else, counted the turning of the seasons by calendar days and holidays and the annual summer vacation from the daily drudgery of school. Since fleeing to this place (via California and

Montana), I have learned to look beyond these human constructs, relying on nature for my seasonal calendar.

Spring begins the day the winter's snow has melted sufficiently to allow me once again to drive the last steep stretch of dirt road up to the cabin, and becomes official when the quakies go to flower and our resident robin pair returns to their nesting territory just up the hill and starts poking about in the moist snowmelt soil for fat, sluggish earthworms.

Summer arrives on the wings of the flaming scarlet and yellow western tanager; the shy, nervous mourning dove; the graceful night-hawk (*Chordeiles minor*) with its darting flight, daring dives, and sonic booms; the mosquitoes upon which the nighthawk feeds; and, most

ASPEN LEAVES — ON THE TREES AND OFF — HELP TO NOURISH SOME THREE HUNDRED SPECIES OF INVERTEBRATES, THUS FACILITATING

THE TRANSITION FROM VEGETABLE TO MEAT AND PROVIDING A STRONG FOUNDATION LINK FOR THE FOREST FOOD CHAIN.

significantly, the greening of the aspens.

The onset of autumn — the most sublime of seasons — is heralded by the ringing, flutelike bugling of rutting bull elk, and painted across the hillsides in gold and crimson by the turning of the aspens.

Finally and inevitably, winter begins with the first major, lasting snowfall that bleaches the mountains of their color, strips the quakies to naked skeletons, puts us back on long skis and clumsy snowshoes again, and erases the game trails that loop like drunken skeins all through this country…game trails, my favorite backcountry byways. 𝕮

Wildlife

3

W I L D L I F E

*Aspen forests are more important to more species
of wildlife than all other trees...combined.*

— J E A N E T T A K . H O D G E S
from *Colorado Outdoors*

T he game trail, ancient and dim but still discernible to an
attentive woodsman's eye, angles down across a steep,
south-facing hill grown thick with aspen, pine, fir, spruce, and scrub
oak. Near the base of the hill, the trail passes close by a huge deformed
ponderosa pine, its thick, twisted base emulating the graceful S-curve of
a swan's neck — a deformity suffered, no doubt, when the tree was
young and slim and impressionable and suffered the misfortune of be-
coming severely bent under a winter's weight of snow. From the swan
tree, the old game trail ambles obliquely on down the hill, snaking
through a dense understory of Gambel oak, chokecherry, bearberry,
wild rose, and other green woody stuff.

At its base, this hill meets a gentler slope that drops off to the west for
a distance of a few hundred yards down to the bottom of a narrow valley
carrying a thin stream — a stream that begins up in the high country of
a place called Missionary Ridge (so named because it reminded some im-
migrant Civil War veteran of the infamous Tennessee battleground).
Here, where south-facing hill meets west-facing slope, the game trail

OPPOSITE: THE MYSTICAL WORLD OF THE
QUAKING ASPEN FOREST INCLUDES A RICH
UNDERSTORY OF FERNS AND FALL-
REDDENED GERANIUMS.

PRECEDING PAGES: BALD EAGLE IN
AUTUMN-BARE PAPER BIRCH BRANCHES.

forks, forks again, and finally dissolves amongst a grove of mixed-age quaking aspens.

As the old trail suggests, this place, this hillside quakie stand, has been a wildlife haven for decades (at least), especially for deer and elk and bears. (Only blacks these emasculated days, though grizzlies roamed here aplenty as recently as fifty years ago. What may have been the last San Juan survivor was killed by an outfitter in 1979. The bear got in a few good licks before it died.)

Everything the animals need is abundant hereabouts: pure, reliable water in the perennial creek just down the slope; a lush understory of brush, tall grasses, and wildflowers in which to hide from danger; the trees themselves to provide thermal cover (shade in summer and shelter from the elements — rain, wind, and snow — in winter); food aplenty, with virtually every understory plant that grows here edible for the herbivorous elk and deer and the omnivorous bears as well; and, as an important bonus for the elk, when times get really tough (as they do around February most every winter), there's the nutritious sprouting tips of Gambel oak and quaking aspen, as well as the green, living inner bark of the latter — wapiti survival rations.

Historically, these large wild animals and many smaller others have visited this idyllic Colorado hillside, this rich grove of aspens, to graze

---

"...the trees themselves provide shade in summer and shelter from the elements — rain, wind, and snow — in winter."

---

or browse, to rest, to mate in autumn, and to give birth come late spring.

This special place — this little hillside aspen grove with the dim and ancient game trail winding down into it — is my front yard.

Evidence of wildlife having used "my" aspen grove abounds: Standing like ivory pillars on either side of my front door are two bone-white, mature quakies both of which boast the time-blackened scars of climbing ursines — black bear tracks ascending the trees. Other aspens nearby wear the distinctive long, curving scars left by porcupines, whose primary food is the bark of pine trees, but who nonetheless occasionally

In winter, starving elk feed on the green inner bark of the aspen, leaving distinct wounds that eventually heal over.

Lacey offered a curious whimper, the elk did not run, nor even bark
(a very loud, doglike sound, which is the universal elk alarm cry). They
know us. They trust us. They have to if they want to utilize our oak
brush and the bark and twigs of our aspens for winter browse.

Although it is far from the top of the elk's preferred foods list, the
soft living bark of aspens is a late-winter staple for the big herbivores
(up to one thousand pounds and occasionally more). It's the nutritious
green inner bark, or cambium, they're after — one of the few moist,
flavorful (for elk) foods available during the frozen months. Each year,
after the snow melts out, during the long forest walks I take to celebrate
the arrival of spring, I find scores — probably hundreds if I were ever to

ABOVE: NORTH AMERICAN ELK, OR
WAPITI, COW AND BULL.

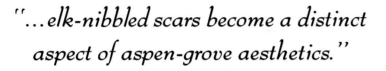

"...elk-nibbled scars become a distinct
aspect of aspen-grove aesthetics."

take the trouble to count them all — of mature aspens, both recently
fallen and standing, that have had their bark chewed by elk.

With living trees, the damage is generally limited to a few parallel
rows of horizontal scars; elk never girdle a live aspen, removing the
bark all the way around and thereby dooming the tree to slow starva-
tion and certain death. With time, the shallow elk-chew wounds
harden, turn black, and puff out slightly, like proud flesh. All is healed.
The only potential damage resulting from elk nibbling on live mature
aspens may arise from the infiltration of decay organisms through the
wounds. Most often, however, no lasting harm is done, and the scars
become a distinct aspect of aspen-grove aesthetics.

(The aspen's soft, dark-scarring, tabula rasa bark attracts not just
antlered and clawed artists, but humans with knives and time on their
hands. An open-eyed walk through any not-too-remote western aspen
grove is likely to reveal carvings that range from simpleminded graffiti
to some delightful aspen art. Much or most of the best was carved in the
early decades of this century by lonely Hispanic sheep herders and is
frequently of a mildly erotic nature.)

With recently fallen live aspens — blowdowns — hungry winter
wapiti use no such restraint and may totally denude the fallen trunk of
its accessible bark.

Come spring, both elk and deer feed heavily on the budding tops of aspen saplings. The quakie is truly a wildlife food for all seasons, just as it has become rich food for thought for me.

Other wild animals, as well, consume the bark, buds, and branch tips of quaking aspens. Even the porcupine — that familiar large prickly rodent generally thought to prefer the bark of conifers — feeds on aspen, consuming the green leaves in summer and the bark in winter. Several years ago, a single hungry porky wiped out an entire little community of aspen saplings growing close by our cabin, stripping every shred of accessible bark from the little trees — from their tips right down to snow level — in a single night. It made me mad.

Then there are the bears. In early spring, soon after emerging from their dens, when the hunger for green stuff is working deep within them, black bears climb mature aspens to break off limbs for their abundant buds and catkins and, a little later, new leaves. When a

''...the mammal most closely associated
with aspens is the beaver...''

sufficient pile of limbs litters the ground around the dinner tree, the bruin descends to feed.

But the mammal most closely associated with aspens — in fact, it has a symbiotic (mutually beneficial) relationship with them — is the beaver, North America's largest rodent. Not only is the aspen (and, to a lesser extent, its relatives in the willow family) the beaver's primary food, it is also the material of choice for building both its wigwamlike homes and the log-and-mud dams that create and retain its watery world.

The average adult beaver will consume a little over four pounds of aspen bark a day — that's some fifteen hundred pounds per year — and prefers to peel its dinner from relatively small-diameter trees and limbs. In the fall, by way of storing up for winter, these rodents are as busy as — well, as beavers — cutting aspen branches (from as far "inland" as three hundred yards and occasionally farther, though most activity is restricted to three hundred feet around the shores) and anchoring them

PRECEDING PAGES: MULE DEER DOE PAUSES IN THE MIDST OF FOREST WILDFLOWERS.

to the bottoms of their ponds. When snow covers the land and surface ice seals the ponds, the big aquatic mammals simply slip into the water through ports in the floors of their lodges and swim down to retrieve an ice-water-preserved dinner.

One study suggests that perhaps two hundred quakie trees of various sizes are required to feed one beaver for a year. That volume of consumption, as you can well imagine, tends to be hard on the local aspen forest.

This all sounds pretty one-way: The beaver cuts the aspen, strips and eats its bark and twigs, then uses the peeled limbs and trunks to build its dams and lodges. So how is the relationship symbiotic? In the words of Jeff Elliott, writing in *Glacial Erratyic*, "Beaver create ponds, which go dry and form meadows, which are transformed into alder runs, which mature into aspen stands...."

In other words, the beaver helps the mature (and, therefore, frequently diseased) aspen grove to regenerate itself. Of course, it takes a good long while — from a dozen or so years up to centuries, depending on the circumstances — for a good crop of aspen saplings to reappear in a beaver-cut area.

ABOVE: BEAVER, BY CREATING PONDS, PROVIDE HOMES AND SUSTENANCE FOR NUMEROUS WILD CREATURES, NOTABLY TROUT AND WATERFOWL.

Those varying circumstances are succinctly described by Norbert V. DeByle, principal plant ecologist at the Forestry Sciences Laboratory at Logan, Utah, in his paper "Animal Impacts [on the aspen ecology]":

> Beaver effects can be placed into two categories: that from cutting alone, and that from dam building and flooding. Cutting alone stimulates abundant suckering [production of saplings]. If beaver abandon that section of the stream for a sufficient time (fifteen or more years) and ungulate use is not excessive, a new stand of aspen will develop. Flooding changes the entire plant community and, to some extent, even the landscape. Siltation behind beaver dams results in a series of benches, each relatively flat and wet (often too wet for aspen to develop), along the stream course. These benches may remain dominated by other vegetation for centuries [before aspens reinvade].

Good, bad, or neutral, the beaver, with its rich, close fur; webbed feet; oversized choppers; big, flat tail; stick dams; stepped pond systems; bank tunnels; mud slides; shaggy lodges like islands of brush; and acres of pencil-pointed aspen stumps, is inseparable in most knowledgeable observers' minds from the quaking aspen forest — and, likewise, the aspen forest from the beaver. For myself, the Rockies just wouldn't be the Rockies were they devoid of their tens of thousands of beaver ponds, the heads of the busy builders occasionally cutting the evening surface as they swim, parting the dark glassy water with ever-widening Vs.

(Even so, I do sometimes wish the possessive beasts weren't quite so persistent in swimming up to evening beaver-pond anglers and repeatedly tail-slapping the water — *Blam! Blam!* — like so many shotgun blasts. For them, the tail slap is merely a territorial warning, but for the angler who may have hiked a mile or more to reach a particularly lucrative brookie pond, suffering swarms of mosquitoes en route, and who intends the beavers there no harm — and no matter how highly he may otherwise think of the slappers — the frustration is keen.)

ABOVE: BEAVER DAMS PUNCTUATE ASPEN FORESTS THROUGHOUT THE ROCKY MOUNTAIN WEST.

OPPOSITE: A SINGLE ADULT BEAVER MAY CUT AND CONSUME THE EDIBLE PORTIONS OF AS MANY AS TWO HUNDRED ASPENS PER YEAR.

Without doubt, the most harmful mammals to the aspen ecology
over the long haul are not beavers, but domestic cattle and sheep,
particularly the latter. According to one scientific study, "When grazed
at similar intensities, sheep were four times more destructive to aspen
suckers [saplings] than cattle."

Concerning the relative harm to the aspen ecology resulting from the
grazing of domestic livestock versus that of wild elk, Norbert DeByle
says that "After the impact of livestock, the additional impact of elk
scattered over their summer range is seldom even measurable."

The management implications are clear: In order to maintain a
healthy aspen ecology that can support itself as well as an abundance
of wildlife, livestock grazing must be strictly limited and carefully moni-
tored. For this reason and many good others (the destruction of ripar-
ian habitat and the consequent watershed damage; severe overgrazing
of grasslands, leading eventually to desertification), many informed and
concerned groups and individuals feel strongly that livestock grazing
should be totally eliminated from public lands. The key players, of
course — those with vested interests in public-lands livestock grazing,
notably the Bureau of Land Management, the U.S. Forest Service, and
the permittee ranchers themselves — are predictably vocal in their
disagreement.

This leads us to a discussion of an odd mammalian species that
uses — and too often abuses — the aspen ecology to make its life a little
easier, a little better...*Homo sapiens.* Of particular interest is the subspecies *H. pyronimrodicus*, the firewood hunter, to which hardy race
I claim active membership.

A Modest Wood

4

A   MODEST   WOOD

There is but one requisite of a fire,
that it should burn.

— DONALD C. PEATTIE
From *An Almanac for Moderns*

A lthough it has its pleasant distractions, "gettin' in your wood," as self-sufficient country folk everywhere refer to the annual task, is nonetheless a job of work. But a most pleasant, satisfying species of work it is. A California new age mystic (or an old age Buddhist monk) would call it karma work, meaning that it warms and exercises the spirit as well as the body. It makes you feel good about yourself. And, too, gettin' in your wood provides workaholics (those of us whose overdeveloped consciences make it difficult to indulge in any task deemed "nonproductive" by Western society) with a superb nonrecreational excuse to spend a lot of time out in the woods, doing something that's not just productive, but truly enjoyable.

Up here in the San Juans, where the wood-burning season extends from mid-September through late May each year — plus several chilly evenings in between — we burn four to five cords of wood per year (a cord being a fairly tight stack measuring eight feet long, four feet wide, and four feet high). That's a lot of wood if you cut, haul, split, and stack it yourself — a lot of wood and a lot of work. And so it is, within

OPPOSITE: SELF-SUFFICIENT FOLK KNOW THAT WOOD WARMS YOU NOT ONLY WHEN BURNING, BUT ALSO IN THE GATHERING.

PRECEDING PAGES: "GETTIN' IN YOUR WOOD" REQUIRES MANY HOURS OF SPLITTING AND STACKING.

73

certain rural social circles, hereabouts as elsewhere, that a person's competence is measured not by the size of his home, the value, newness, or nationality of his vehicle, but by the size of his self-got autumn wood pile. "Got your wood in yet?" is a standard September-October greeting amongst *Homo pyronimrodicus*; the question would never be asked of one who buys his wood, for its answer would imply nothing in the way of physical effort or spiritual resolve.

It has been said, and often repeated, that firewood warms you twice — first when you gather it, then again when you burn it. Well, yes, that it does, at the very least. As one who has heated exclusively with wood for a decade now — no cheating by buying it cut, split, delivered, and stacked and no fossil-fuel or electric appliances to fall back on should the wood pile prove too small along about February or March — I reckon that gettin' in your wood warms a body more like half a dozen times.

First and most invigorating, gettin' in your wood warms you when you enter the forest, chain saw in hand (oh, those cranky, snarling,

"...hereabouts, a person's competence is measured by the size of his self-got autumn wood pile."

potentially hurtful little devils!), locate a deadfallen or standing dead tree ("snag"), check to make sure it's not hollow or broken-topped and therefore likely to be some wild creature's winter home, fell it, limb it, and cut ("buck") it into manageable logs (their lengths varying according to the diameter and weight of the wood), drag, carry, or roll each length to your truck or utility trailer, and wrestle it in.

Wood warms you a second time when you get home and unload your logs, either stacking them neatly to make the next chore simpler, or merely tossing them to the ground in a shaggy pile.

Wood warms you a third time when you saw the logs into stove lengths ("rounds") of twelve to twenty inches each, depending on the size of your stove or fireplace.

Fourth, when you swing axe or maul to cleave each round, according

SEPTEMBER SNOW SQUALL, SAN JUAN MOUNTAINS, WESTERN COLORADO.

to its diameter, into two, four, or more splits.

Fifth, when you haul those splits — by armload, wheelbarrow, or vehicle — to their final resting place near (but, due to fire hazard, not smack against) your abode and stack them, bark-side up (the bark acts as a roof to help shed water) in a neat, stable pile that allows air to

"...gettin' in your wood is most definitely the very best sort of work..."

circulate throughout to promote drying and seasoning. (Cap the top of the pile with a plastic tarp, but — at least until it starts snowing — leave the sides alfresco.)

Sixth, finally, and perhaps most satisfying, gettin' in the wood warms you when you duck out the door on a white winter's morn, grab an armload from the woodpile, dart back inside, feed a couple of splits into the black hungry mouth of the stove, and drop the rest into the wood box for convenient later use.

Just so.

If I don't have to carry it too far, I can comfortably cut and load about half a cord of wood into the bed of my old half-ton pickup in a couple of hours of grunt-work. Thus, to get in five cords of wood, I can look forward to ten wood-gathering trips to the forest between spring and fall, along with ten repetitions of each of the other five steps. Plus, of course, a myriad of peripheral but nonetheless essential chores: cleaning and tuning the chain saw (that, I must admit, gets old), sharpening the saw chain (ditto), sharpening the splitting maul and hand axe, making and stacking kindling, cleaning up and hauling away or burning the small mountain of debris generated by cutting and splitting several cords of wood, emptying the stove's ash pan (a daily chore in winter), cleaning the stovepipe (up on the roof, down on your knees), and more.

Yes, gettin' in your wood is most definitely a job of work. But it's the very best sort of work, its rewards far more tangible, satisfying, and lasting than an ephemeral paycheck. In fact, once autumn arrives and we fire up the stove for another long season's use, the payoff comes daily.

OPPOSITE: WOOD-BURNING SEASON BEGINS SHORTLY AFTER THE FIRST AUTUMN FROST AND LASTS WELL INTO APRIL.

FOLLOWING PAGES: AUTUMN ASPENS CANOPY A YOUNG EVERGREEN.

The little cabin was plenty cold this morning, as it is every fall morning up here in the aspen forest. But within seconds of crawling from bed, I had a fire roaring in the trusty wood stove, "snapping and crackling and promising life with every dancing flame," as Jack London so poetically put it.

For years, our heater was an ancient cast-iron potbelly of Monkey Wards lineage, which I rescued from a demolished ranch house a few

miles down the valley. A couple of winters back, we moved up to a more efficient soapstone airtight model. With both species of wood-burner, and most others, the firing routine is simple and dependable.

On all but the coldest of nights, I allow the fire to die down in the wee hours (largely because I'm too zonked to get up and stoke it). Come morning, I crumple a few sheets of newspaper and toss them into the empty belly of the stove. Over the paper I arrange a tipi of half a dozen finger-thick kindling sticks. Over these I stack a second tipi of three or four arm-sized splits of seasoned firewood. Next, I open both the stovepipe damper and the spin-draft in the ash-pan (bottom) door, then fill the blue-speckled cowboy kettle with well water and replace it on the stove top.

Now, all that's left is to strike a match, touch it to the crumpled newspaper at the bottom of the fuel pile (or simply stir the coals if any remain glowing from the night before, which, with an efficient stove, they should), close the door, and I'm on my way to comfort and coffee.

And far more often than not, the wood that I fell, haul, split, stack, and burn — as it is for many another westerner who gets in his own — is what an elderly lady rancher friend lovingly refers to as "that good ol' aspen wood."

ABOVE: AMONG THE SOFTEST OF THE HARDWOODS, QUAKING ASPEN CUTS AND SPLITS EASILY.

Most books and articles on wood burning, written in the East with an eastern bias, don't even mention aspen when rating fuelwoods. Those few that do invariably place it near the bottom of the list for efficiency and, it is implied, for overall desirability.

For an instance, Maine's John Vivian, a respected back-to-the-land self-sufficiency authority and author of the fine little book *Wood Heat*, put it this way in a booklet he wrote for the Stihl Chain Saw Company: "And then there's 'gopher wood' such as alder, aspen, or cottonwood; fill the firebox and you have to turn around and 'go fer' another load, it burns up so quickly."

Although Vivian indulges in a bit of hyperbole there, his point is valid; it makes good quantitative, technical sense, since aspen (like cottonwood, which is also a member of the willow family, you'll recall) is among the softest of the hardwoods. Soft suggests a lack of density and, in turn, light weight when dry, yielding a rapid burn and a low

"The little cabin was plenty cold this morning, but within seconds I had a fire roaring in the trusty wood stove."

Btu-per-cord heat output relative to harder, more dense and heavy woods.

For those who buy their fuelwood, that's important to know, since the less efficient a particular species of wood, the less you should pay for it. Here in southwest Colorado in the winter of 1990–91, for example, a cord of aspen, split and delivered locally, costs about seventy dollars. Ponderosa pine is about the same, while Gambel oak — a superb stove fuel — goes for one hundred twenty-five dollars.

Now, if I were buying firewood, I'd choose aspen over the equivalent-priced pine, since the heat output of the two is roughly the same and aspen is better-mannered as regards cleanliness (that rough pine bark can make a real mess in the wood box and on your floor), sparking, odor, and creosote production (about which more in a moment). Yet, I'd choose oak over either aspen or pine, since a single cord of oak will yield almost as much heat as two cords of aspen or pine while offering the significant benefit of having to handle only half as much wood, load the stove and empty the ash pan half as often. (Actually, the perfect combination is about fifty-fifty, using aspen for kindling and quick starts, and oak for the long, slow burn.)

To quantify that comparison, a cord (about eighty-five cubic feet) of seasoned oak will yield close to twenty-three million Btu. As you may (or may not) recall from your school days (yawn), a Btu, or British thermal unit, is the amount of heat required to raise the temperature of one pound of water (about a pint) one degree Fahrenheit. Meanwhile, a cord of dry pine, depending on the species, puts out thirteen to fifteen million Btu, and a cord of seasoned aspen a little better than fourteen and a half million Btu.

Thus, the telling measurement is not volume of wood as reckoned in cords, but dry weight. While a dry cord of Gambel oak might weigh close to three thousand pounds, the same volume of dry aspen would go perhaps only a third of that.

Again, this is important information for those who purchase fuel-wood. But for those of us who get in our own, such technical ratings don't tell the full story. In fact, they're hardly worth the paper they're printed on. (Which paper, incidentally, if it's of the finest quality, stands a fair chance of having been manufactured from aspen pulp.) Btu-per-volume ratings constitute merely one of the several factors used in calculating an intelligent personal fuelwood preference. Other, equally important considerations include:

Ease of finding and gathering. With most species of western trees, the wood-gatherer has to search far and wide, locating a snag here, a deadfall there. This requires lots of driving, lots of walking, lots of hauling, lots of energy and time. But since aspen tends to grow in dense groves and is relatively short-lived, where you find a little you'll often find a lot. A single small aspen grove, unpicked-over by previous gatherers that season, will typically contain a dozen or more deadfalls, a dozen or more snags, and an equal number of obviously ill trees. Given that a single mature aspen will yield approximately one-quarter cord of cut, split, and stacked wood, you can often harvest a full winter's supply from a single grove. (And, as a spin-off benefit, due to a natural phenomenon explored in a later chapter, selectively cutting ailing aspen actually improves the health of the grove and speeds regeneration.)

Ease of cutting. The denser and harder a wood, the longer it takes to make each cut through it and the quicker your saw chain will dull in the doing. Thus, all things considered, in the time (and with the energy) it takes to cut one cord of oak, I can cut at least two and perhaps as many as three cords of the larger, softer, and less-limby aspen.

OPPOSITE: SINCE ASPEN TEND TO GROW IN DENSE GROVES AND ARE RELATIVELY SHORT-LIVED, WHERE YOU FIND A LITTLE YOU'LL OFTEN FIND A LOT.

■ The Forest Service claims that many aspen forests need to be clear-cut in order to save them from invasion by conifer forests. The conservation groups have long criticized the agency for this assertion, *which has been called into question by the agency's own research stations.*

■ The Forest Service's preferred alternative calls for twenty two miles of new local road construction *per year.* Road density can affect recreation experiences and big game habitat quality.

The conservation groups argue that public tax dollars should not be spent to prop up L-P when the private market has already shown that *the highest value of aspen trees is not their potential for timber, but rather their value for supplying wildlife, grazing and recreation.*

"The Colorado Wildlife Federation is disturbed," said Executive Director Steve Blomeke. "We understand that when L-P first came into the state, Regional Forester Craig Rupp told the company that the Forest Service would allow cutting of aspen forests for perpetuation of the species and wildlife habitat, not for fiber production. L-P agreed to that. Now, the Forest Service seems to have reneged on this agreement."

And so on.

Understandably, L-P's only allies in this brouhaha are those who depend, directly or indirectly, on the mill's continued — and, if possible, ever-expanding — operation for their prolonged profitable tenure here in beautiful but economically depressed western Colorado. Included among this self-serving number, it seems, is the U.S. Forest Service. Why else would the "Freddies" (as one outspoken environmental group is wont to call them, no compliment intended) have us believe that the best way to "save" an aspen forest is to cut it down? In-credible.

So much for the seamy side of the otherwise idyllic world of the quaking aspen. I'd rather talk about — of all things — sex. ❦

Life
Renewed

6

L I F E   R E N E W E D

Ecology, by its essence, is a science of digressions
and footnotes and parentheses.

— DAVID QUAMMEN

from *Outside*

Quaking aspen trees — and, on the larger scale, quaking aspen groves — reproduce in two distinct fashions: through flowering and the consequent manufacture and distribution of seeds; and vegetatively, by genetic cloning to produce root suckers (saplings that sprout directly from the roots of parent trees), effectively bypassing the more risky seeding process. This reproductive duality exists not as a result of evolutionary caprice but for a variety of good and time-proven reasons.

Aspens normally produce seeds in profusion. However, the environmental requirements for the successful germination and survival of these seeds are so very stringent as to severely limit the effectiveness of this most traditional method of sylvan reproduction. Still, each and every year, just as it has been for untold millennia, the tenacious quakies perform the same sacred dance.

The first signs are so subtle as to go unnoticed by the casual eye. Indeed, even to the interested observer standing within a grove beneath the trees, the changes are all but invisible. But from afar, all is revealed: By midwinter each year, the tops of the collective thousands of skeletal

OPPOSITE: STRONG AND HEALTHY ASPENS WILL GROW NEARLY A FOOT TALLER EACH YEAR.

PRECEDING PAGES: AN "AVERAGE" ASPEN MAY PRODUCE FROM TWO HUNDRED THOUSAND TO TWO MILLION SEEDS, YET FOR ALL OF THIS HARD AND CREATIVE WORK, IT ISN'T AT ALL UNUSUAL FOR NOT A SINGLE SEED FROM AN ENTIRE GROVE TO SPROUT AND GROW TO MATURITY.

aspens that cling to the steep slopes across the little river valley above which I live — as elsewhere throughout aspen country — have taken on a dull but distinct rust-red, the first visible signs of budding. Actually, the budding process begins in late summer/early fall, but only with subsequent swelling and reddening during winter dormancy does it become obvious.

Why so early? Because, in order to germinate, aspen buds have to undergo several weeks of subfreezing temperatures. No matter: By midwinter, craving as I invariably am for warm weather, I choose to take this blushing of the aspens as a portent of spring pending — although Easter and the equinox are yet several long weeks away.

By late winter/early spring, the small, tight, rosy aspen buds have become individually distinct, with those growing low enough to be accessible providing a favored food for winter-starved herbivorous wildlife. By late April or early May, the buds have opened and their flowers appeared. Even though aspen "flowers" are not what most would call beautiful, nor even properly florid, their annual appearance is spiritually refreshing and always welcome.

As with most florescent plants, a given aspen flower will normally be either male (staminate) or female (pistillate), the two sexes generally

> "...in order to germinate, aspen buds have to undergo several weeks of subfreezing temperatures."

occurring in a convenient one-to-one ratio among the trees within a grove. Departing, however, from the floral norm, a single aspen flower does not usually produce both stamen and pistil (those within a given grove that do may range from 5 to 20 percent).

When the time is ripe, a plenitude of yellow pollen is released from the anthers of the staminate flowers and distributed far and wide by winds and breezes. At this time of year, here amidst our aspen grove, everything for a while is yellow with pollen dust: yellow on our cabin's brown metal snow roof, yellow on our old red-and-white pickup truck, yellow dust in the road, yellow rings around evaporating rainwater

Winter

7

WINTER

*In my imagination, through ways incomprehensible
to the author, desire and love and death lead through the
wilderness of human life into the wilderness of the natural
world—and continue, round and round, perhaps forever,
back again to wherever it is we began.*

— EDWARD ABBEY
from the Preface to *Black Sun*

F ebruary: The shortest and darkest of months.

It has been ten weeks now since the annual Thanksgiving
snowstorm — a remarkably predictable meteorological event here in the
southern Rocky Mountains — blew in to bury the steep, winding dirt
roads that loop through this obtrusive little settlement. Although it may
sound foreboding, this is good for Carolyn and me: As the snows arrive
each year, so depart the last straggling few fair-weather summer-home
vacation residents, packing hurriedly, tacking sheets of unpainted
plywood over the windows of their unwinterized prefab cabins, scurry-
ing down and out of the mountains and on south and west to Sun City
and similar winter havens.

Behind them they leave tranquillity, snow-closed roads, and a thankful handful of year-round residents. Admittedly selfish about my soli-
tude (it's *so* hard to come by these days), no one is more thankful to see
them go than I. The fact that some of them are good, even enjoyable,
people matters not one whit. That isn't the point.

A meek sun rides low along the southern sky these brief winter

OPPOSITE: EARLY WINTER STORMS STRIP
THE QUAKIES OF ANY REMAINING FOLIAGE,
LEAVING ONLY NAKED SKELETONS.

PRECEDING PAGES: WINTER BEGINS WITH
THE FIRST MAJOR, LASTING SNOWFALL
THAT BLEACHES THE MOUNTAINS OF THEIR
COLOR. LA PLATA MOUNTAINS, SOUTH-
WESTERN COLORADO.

days — that is, when it isn't obscured entirely by the ghostly fog of a total whiteout. In such conditions — white above, white below, white all around — the barren aspens seem gray, sad, pensive, humble. Or perhaps I am merely projecting my own melancholy upon them.

February, the darkest and shortest of months. But hardly short enough.

The much-heralded and at first welcome "winter wonderland" of Christmas celebration, given a few weeks to slip into retrospect, loses much of its charm, becomes a drudge. Autumn seems a mere faint memory, spring a hopeless ambition. Cabin fever is endemic. Put another log on the fire please Mama, boil us up another pot of tea. Or better yet, a pot of strong black coffee made from stout, aromatic beans grown someplace far to the south, a place where bright colorful birds sing from verdant forests and February is among the most pleasant of months.

But even in February, even up here in the frozen white Rockies, life goes on. Especially around the slender aspen from which hangs Carolyn's bird feeder. To this winter oasis flock not just birds but also the big gray and white tassel-eared Abert squirrel (*Scircus abertii*); the occasional lean cottontail or plump snowshoe hare; and, keeping watch with winter-hungry eyes from aeries atop nearby towering spruce, fir, or ponderosa, the winged predators — hawks, owls, falcons. These skilled

ABOVE: JUNCO, ONE OF SEVERAL HARDY SMALL BIRDS THAT MAKE THEIR YEAR-ROUND HOMES AMONG THE ASPEN.

> *"…in February, even up*
> *here in the frozen white Rockies,*
> *life goes on."*

hunters crave the warm moist flesh of the raucous Steller's jays, the ubiquitous junkoes, the diminutive, sweet-voiced nuthatches and chickadees that frequent the aspen feeder.

One February morning last winter, as I stood at the kitchen window sipping coffee and watching the big blue Steller's squabbling over seed (needlessly, since I had only minutes before filled the feeder and there was plenty for all), a big, dark shadow dropped from above to knock one of the robin-sized jays from the feeder perch, pinning it to the ground directly below.

MULE DEER SKULL AMONGST FALLEN AUTUMN ASPEN LEAVES.

Book Design by Larry Lindahl

Calligraphy by John Fortune

Typeset in Bernhard and Newtext by ProType, Phoenix

Printing by Lammar Offset Printers